How Animals Talk

by Susan McGrath

Japanese cranes call and dance together.

■ BOOKS FOR YOUNG EXPLORERS
NATIONAL GEOGRAPHIC SOCIETY

Wolves howl and yap and growl. A coyote lifts its head and howls. Wolves and coyotes don't use words. But they do send messages with sounds and smells and in other ways. All animals tell each other things. They communicate.

An angry wolf raises its back and walks with its legs stiff. With its body, the wolf shows other wolves what it is feeling.

A wolf sniffs the snow
for smells left by other wolves.
A male deer rubs against a tree,
leaving his smell there.
The smell tells other deer,
"Stay away from here."

Animals don't talk as we do.
But they have many different ways
of communicating with each other.

Three white-tailed deer flick their tails up as they race across a stream. The snowy white tails alert other deer to danger. The signal means, "Run! Follow me!"

A little pika communicates danger, too. It calls out a loud warning, "Eeek!" Other pikas hurry to safety.

Crack! Two bull elk crash horns. "Which one of us is the boss?" they communicate to each other. "Let's see which one is stronger."

Two arctic hares send the same message. The large, woolly hares jump up and hit at each other with their paws.

Insects also communicate.
A praying mantis makes itself
look as big as it can.
This is a warning that says,
"Better leave me alone."

Even a caterpillar has things to say.
This swallowtail caterpillar
can give off a bad smell that
helps keep enemies away.

The light of the male firefly says,
"Here I am." To a female passing by,
his signal says, "Come over here."

The male spider, the smaller of
these two, walks carefully
on the female spider's web.
"I am your kind of spider—
not your food," his steps say.
Then, she lets him come in.
The red shape on her belly
tells us that she is
a poisonous black widow spider.

14

ROBINS

Three baby robins open their mouths wide for a meal of worms. "Feed me," the open mouths signal. "Feed me! Feed me!"

Barn swallows chatter on a branch. Chinstrap penguins seem to have a lot to say. But no one knows just what the birds are saying. We can only guess. What do you think they are talking about?

BARN SWALLOWS

CHINSTRAP PENGUINS

15

Big birds called albatrosses
are getting to know each other.
They clap their bills and move
around. These birds are also
called gooney birds.

A male gooney bird dances
for his mate. He twists his wings
and bows his head. "Will you
be my mate?" his dance says.

"Yes, I'll be your mate,"
she lets him know.
After a while, he hugs her
with his neck.

At nesting time, the male
frigate bird puts on a show.
He puffs up his throat sac
like a bright red balloon.
He flies high, showing off
his beautiful colors. "Look at me,"
the male signals the female.
"Won't you be my mate?"

At other times, the male
lets the air out of his throat sac.
The birds rest on their nest of twigs.

Blue-footed boobies dance
and pose for their mates.
One part of the dance is called
"sky pointing." Can you guess why?

Like the frigate bird, the booby uses
his body to show off for his mate.
He parades up and down,
showing off his bright blue feet.
He whistles softly.
"Ark, ark," she answers.

Two male elephant seals rise up in a crowd
of females. The males, called bulls,
snort and roar through their big noses.

"This group is mine!" one bull roars.
"No, no. It is mine!" the other roars.
The females don't seem to notice all the noise.

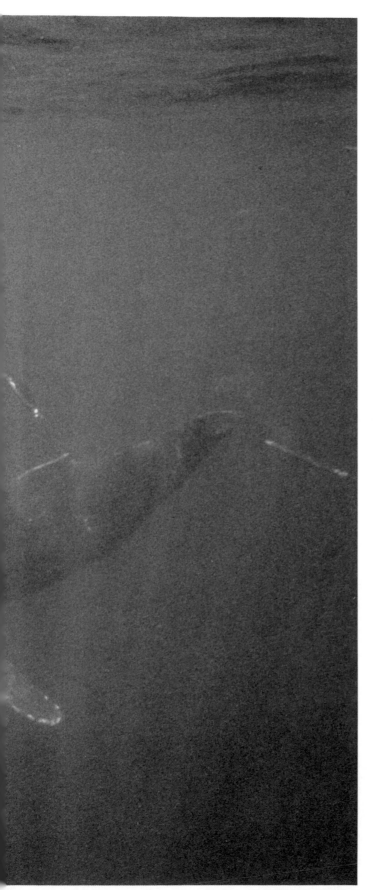

Even under the sea, animals communicate with each other. Three humpback whales— a baby, a female, and a male— are swimming along together. The male sings a loud song. "Here I am," says his song.

A beluga whale chirps and clicks. Its sounds made sailors think of a songbird. They used to call the beluga "the canary of the sea."

BELUGA WHALE

HUMPBACK WHALES

Two sea mammals called manatees meet and touch whiskers underwater. Manatees often kiss like this when they greet each other. They also chirp and squeal.

Who ever would have thought that animals could have so much to say! On the ground, in the air, and under the water, they are sending signals.

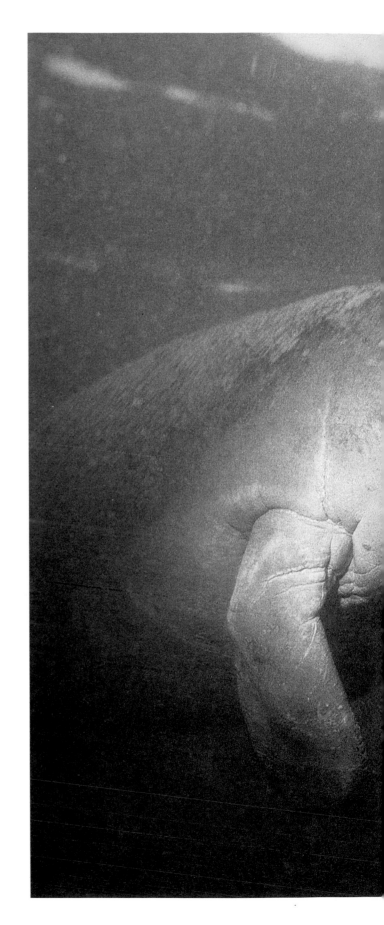

An older female grooms the hair of a young chimpanzee. She picks out pieces of dry skin with her fingers. Chimpanzees often comfort each other by grooming. "I like you," their touch says.

A young owl pecks an adult, begging to be fed. All animals communicate, using sight, sound, touch, and smell. Next time you see birds, or squirrels, or cats, or dogs, watch them closely. What do they have to say?

A young prairie dog
seems to be whispering
secrets to its mother.
Touching and sniffing one
another helps prairie dogs
get along together.

Cover: A male red deer
gives a trumpeting roar.
This loud call lets
other red deer know
that he is there.

Published by
The National Geographic Society, Washington, D. C.
Gilbert M. Grosvenor, *President and Chairman of the Board*
Melvin M. Payne, *Chairman Emeritus*
Owen R. Anderson, *Executive Vice President*
Robert L. Breeden, *Senior Vice President,*
Publications and Educational Media

Prepared by
The Special Publications and School Services Division
Donald J. Crump, *Director*
Philip B. Silcott, *Associate Director*
Bonnie S. Lawrence, *Assistant Director*

Staff for this book
Jane H. Buxton, *Managing Editor*
John G. Agnone, *Illustrations Editor*
Lynette R. Ruschak, *Art Director*
Peggy D. Winston, *Researcher*
Sharon Kocsis Berry, *Illustrations Assistant*
Carol R. Curtis, Mary Elizabeth Ellison, Rosamund Garner,
Bridget A. Johnson, Artemis S. Lampathakis, Sandra F.
Lotterman, Eliza C. Morton, Virginia A. Williams,
Staff Assistants

Engraving, Printing, and Product Manufacture
Robert W. Messer, *Manager*
George V. White, *Assistant Manager*
David V. Showers, *Production Manager*
George J. Zeller, Jr., *Production Project Manager*
Gregory Storer, *Senior Assistant Production Manager*
Mark R. Dunlevy, *Assistant Production Manager*
Timothy H. Ewing, *Production Assistant*

Consultants
Eirik A. T. Blom, Maryland Ornithological Society; Sally
Love, Smithsonian Institution; Craig Phillips, Biologist;
William A. Xanten, Jr., National Zoological Park,
Smithsonian Institution, *Scientific Consultants*
Peter L. Munroe, Dr. Ine Noe, Karen O. Strimple,
Educational Consultants
Dr. Lynda Bush, *Reading Consultant*

Illustrations Credits
Manfred Danegger/PETER ARNOLD, INC. (cover); Steven C.
Kaufman/PETER ARNOLD, INC. (1); Jim Brandenburg (2-3, 4
left); Jerry L. Ferrara (3 upper right, 31 right); Thomas Kitchin (3
lower right, 7 right, 8-9); Leonard Lee Rue III (4-5); Murray
O'Neill/VALAN (6-7); Art Wolfe (9 right, 15 lower); Dwight R.
Kuhn (10-11, 14); C. Andrew Henley/LARUS (11 right); E. R.
Degginger (12 left); James H. Robinson (12-13); ANIMALS ANI-
MALS/Terry G. Murphy (15 upper); Frans Lanting (16-17, 18-19,
19 right, 21 lower right, 23 right); Laura Riley (20-21); C. C.
Lockwood (21 upper right); Gerald & Buff Corsi (22-23); Jeff Foott
(24-25, 27 right, 28-29, 32); Flip Nicklin/NICKLIN & ASSO-
CIATES (26-27); ANIMALS ANIMALS/Zig Leszczynski (30-31).

Library of Congress CIP Data
McGrath, Susan, 1955-
How animals talk.
(Books for young explorers)
Bibliography: p.
Summary: Describes how animals communicate with each
other by means of sight, sound, smell, and touch.
1. Animal communication—Juvenile literature. [1. Animal
communication. 2. Animals—Habits and behavior]
I. Title. II. National Geographic Society (U.S.) III. Series.
QL776.M39 1987   591.59   87-14173
ISBN 0-87044-679-7 (regular edition)
ISBN 0-87044-684-3 (library edition)

MORE ABOUT How Animals Talk

Almost every animal in the world communicates with others of its kind. Communicating helps animals find food and avoid danger. It helps them live together in relative harmony. Even more important, communicating helps animals attract mates and raise their young. In this way, they guarantee that their species will continue.

In the diverse animal kingdom, creatures have evolved a variety of ways to get messages across—by barking and howling, by leaving a scent trail, by touching noses, by flashing a light in a specific pattern.

Though other animals don't use words, as humans do, they communicate with signals that other creatures can see, hear, feel, or smell. Animals may combine different means of communicating, employing more than one sense.

Wolves, for example, display a variety of behaviors, conveying many different messages. These mammals live in packs, usually composed of two parents, their cubs, and other adults. The size of a pack can vary greatly, but most consist of about eight wolves. There is a well-defined social order, or dominance hierarchy, among wolves. One male and one female are the top-ranking members. The male is in charge of important activities such as traveling and hunting.

Often, when two wolves meet, the dominant male raises his tail and walks stiffly. The subordinate male crouches, with his head and tail lowered and his ears flattened. He licks the leader's muzzle and whines softly. With visual display (his pos-

ture), touch (the lick), and sound (the whine), the wolf is signaling acceptance of the other's superiority, thus avoiding a fight.

Wolves hunt in packs for large prey such as deer. Before the hunt begins, the wolves gather at dusk for a group howl (2-3).* Howling helps them keep in touch and announce their presence to other wolves. Wolf packs occupy a territory, the area in which they hunt and raise their young. They advertise their boundaries to neighbors by leaving their scent on trees and rocks, as well as by howling.

A male white-tailed deer also leaves scent marks, particularly during the breeding season, using

*Numbers in parentheses refer to pages in *How Animals Talk.*

secretions from glands on his head (4-5). When he rubs his head and antlers against a tree, the deer leaves scent and scratch marks that identify him to others.

Identifying individuals of the same species is an important aspect of communication. There are several species of fireflies, or lightning bugs. How do they tell each other apart at mating time? Each species has its own characteristic flashing pattern. Fireflies are genetically programmed to respond only to the pattern of their own species (12).

The male black widow spider plucks his message on the female's web for a similar reason (12-13). The message identifies him as a

JIM BRANDENBURG

Licking a more dominant wolf's muzzle, a subordinate wolf sends the message: "You're the boss." A dog often behaves this way with other dogs.

male black widow spider rather than as a predator—or prey. He is about half the size of the female, and she is smaller than a dime. The ability to communicate, through the vibrations of his steps as he moves over the web, may save him from being eaten by her.

The courtship displays of birds serve more than one purpose. A male frigate bird's showy red throat sac attracts females (20-21). The display also lets other males know that a territory is taken. An exceptionally graceful flier, the frigate bird nests on islands in tropical regions of the world.

Japanese red-crowned cranes, nearly five feet tall, perform a ceremonial courtship duet and dance (1). The duet, or unison call, tells other cranes to go away. One crane calls, "Doooo," and the other replies, "Doot-doot." These birds usually mate for life and may live more than 50 years.

Blue-footed boobies also carry out an elaborate ritual during the mating season (22-23). Boobies are usually silent, but on the breeding grounds they make trumpeting and whistling noises. Parading and foot raising are accompanied by wing rattling and sky pointing. Displaying for each other helps to reinforce the bond between mated birds. These large Pacific seabirds probably got their name from early Spanish sailors who called them foolish, or *bobos*, because they let themselves be caught so easily.

The Laysan albatross (16-17, 18-19), a large bird that nests on Pacific islands, is sometimes called the gooney bird, probably because of its awkwardness on land. Superstitious sailors once believed that killing an albatross brought bad luck.

Elephant seals (24-25) get their name from their massive trunklike

LEONARD LEE RUE III

An orangutan makes a "pout face," showing anxiety or frustration.

noses, which reach lengths of 15 inches in mature males. Their noses enable them to roar impressively. The sound can be heard half a mile away. During the breeding season, their roaring warns away rivals. The seals establish dominance through threats and fighting. Elephant seals live in the Pacific Ocean from Alaska to Mexico. Though they come ashore during the breeding season, elephant seals spend most of the year out at sea.

Because visibility underwater is usually very poor, sea mammals such as seals, whales, and manatees communicate mostly by sound and touch. The songs of humpback whales—eerie moans and cries, snores and groans—carry through the water for miles (26-27).

Manatees make various noises—squeaks, squeals, and chirps. They nuzzle and nibble each other when they meet, probably as a means of recognition (28-29). These large, plant-eating mammals, also known as sea cows, are exceptionally gen-

tle. They inhabit warm, quiet waters from Florida to northern Brazil.

Grooming is another means of communication. It helps strengthen ties between animals that live together. Apes and monkeys pick through each other's hair with their fingers (30-31). Birds preen one another's feathers with their bills.

To help your Young Explorer understand different ways animals communicate, try doing some people watching together. You might point out that we use more than words. We communicate through gestures, facial expressions, and posture—body language. What examples do you see?

Whether children live in the city or in the country, they can enjoy observing animals communicating. Take a walk together, keeping your eyes and ears open. You can see dogs establishing dominance and marking their territories just as wolves do. They also bow down as an invitation to play.

You're sure to see and hear songbirds announcing their territories. By simply sitting on a park bench, you can watch pigeons' elaborate courtship displays. However, you may never know exactly what the animals are saying to each other.

ADDITIONAL READING

Animal Communication, by Hubert and Mable Frings. (Norman, University of Oklahoma Press, 1977). Family reading.

Animal Language, by Michael Bright. (London, British Broadcasting Corporation, 1984). Family reading.

In NATIONAL GEOGRAPHIC: "The Intimate Sense of Smell," by Boyd Gibbons, September 1986.